# The Bluffer's®
## Guide to
# The Cosmos

Daniel Hudon

Oval Books

Published by Oval Books
5 St John's Buildings
Canterbury Crescent
London SW9 7QH
United Kingdom

Telephone: +44 (0)20 7733 8585
Fax: +44 (0)20 7733 8544
E-mail: info@ovalbooks.com
Web site: www.ovalbooks.com

Editor – Saffron Beatson
Series Editor – Anne Tauté

Cover designer – Vicki Towers
Cover image – © SPL/Photo Researchers, Inc.
Illustration of Solar System: © Fotosearch/Unlimited Images, Inc
Printer – J F Print Ltd, Sparkford, Somerset
Producer – Oval Projects Ltd.

The Bluffer's® Guides series is based
on an original idea by Peter Wolfe.

The Bluffer's Guide®, The Bluffer's Guides®,
Bluffer's®, and Bluff Your Way® are
Registered Trademarks.

Cover: Astronaut floating free during a space walk
or extravehicular activity (EVA), over Earth.

ISBN: 978-1-903096-42-0

# CONTENTS

# INTRODUCTION

You'll get a big bang out of bluffing about the cosmos. After all, there's nothing bigger than the universe. And bluffing about the cosmos can be as easy as time travel once you hurtle a few choice words into the empty space of conversation. It's one of man's most ancient pastimes: even when humans were still living in caves they were already bluffing about whether the patterns of lights in the sky represented the gods and goddesses on holiday or were chance alignments of distant suns undergoing thermo-nuclear fusion reactions in their cores.

Bending the inner space of the minds of your friends comes with the territory. You will be able to talk about concepts like exploding stars that blast you out of your everyday concerns of shopping and sob stories into the realm of celestial delights. And then there's the numbers that are so large they are considered astronomical (from the Latin, *astro*, which means "What's your star sign", and nomical, which rhymes with 'comical'). You should be rubbing your hands in glee.

**❝ The best part of bluffing about the cosmos is that there is always something new under the Sun. ❞**

The best part of bluffing about the cosmos is that there is always something new under the Sun. So if you feel that cosmic microwave background fluctuations are no longer cool, you can easily switch to evaporating black holes, millisecond pulsars or the mysterious dark energy.

> **❝ If you feel that cosmic microwave background fluctuations are no longer cool, you can easily switch to evaporating black holes. ❞**

And if you're ever stuck, you can always quote the bluffer's best friend, ol' Shakespeare himself: "There are more things in heaven and earth, Horatio, than are dreamt of in your philosophy." That is, if you happen to be talking with someone named Horatio.

## THE BIG PICTURE

Your bluffing career in the realm of the cosmos will be greatly enhanced if you get a few things in order first. The terms cosmos and universe are basically interchangeable though 'cosmos' implies order and harmony so you can use this if you're feeling poetic. Astronomers describe the universe as the totality of things, including space and time, which doesn't leave much out, except perhaps your imagination.

Specifically, the universe contains, in order of increasing size, atoms, molecules, dust*, lost sets of keys, asteroids, comets, moons, politicians' egos, planets, nebulae (clouds of gas and dust), stars, galaxies, voids and clusters of galaxies. It also contains at least 6½ billion bluffers, though, in the interests of space, we won't name them here.

A vital thing to get right, even if you can't remember much else, is what zooms around what. So:

- moons circle around or, rather, 'orbit' planets
- which together orbit stars, and
- stars, together with nebulae, dust and **dark matter** orbit their galactic centres, and
- galaxies orbit each other, if they orbit anything at all.

At present, bluffers are known to live only on one planet, Earth, which has a moon with the singularly unimaginative name 'the Moon', and belongs to a clan of eight planets that orbit the Sun along with a myriad comets (giant snowballs), and asteroids, chunks of rock which, due to their inability to navigate in any way besides auto-pilot, occasionally wreak havoc on any planets that happen to be in

> **❝A vital thing to get right, even if you can't remember much else, is what zooms around what. ❞**

their way. This family (a.k.a. the Solar System) and all the stars seen in the night sky live contentedly within the arms of a galaxy which is known by the

---

\* Cosmic dust, small particles of solid matter, much smaller but even more common than house dust. Unfortunately, the vacuum of space won't take care of the problem.

surprisingly soppy name of 'the Milky Way', one of about 100 billion galaxies living in a happiness of truly universal proportions.

## Size matters

When it comes to the size of the cosmos, just remember that it's not size but how you bluff it that matters. True bluffers won't bother about the size of the Solar System. It's too small. Things are always colliding with each other. Even today, five billion years after the Solar System first whirled into existence out of its primordial cereal box, comets, asteroids and meteors are still crashing into planets and moons.

Look at the Moon. Its pock-marked face has been smacked by everything from specks of dust to behemoth boulders. This wouldn't happen if the Solar System lived in a bigger house.

> **❝ True bluffers won't bother about the size of the Solar System. It's too small. Things are always colliding with each other. ❞**

Even Jupiter, a mere five times further away from the Sun than the Earth, a distance you could walk in perhaps 35 minutes (assuming you can walk as fast as the speed of light), was bashed by Comet Shoemaker-Levy not that long ago. Bruised for a few days, the giant planet shrugged it off as if it happens all the time.

There is clearly not enough space in the Solar System.

## The nearest stars

So, bluffing about the size of the Solar System can be done for practice, perhaps as a warm-up, but never as your main attraction. The distances to the stars are a greater challenge and slightly more respectable.

**66 Be advised as you wander the deeps of interstellar space, the nights will seem very long. 99**

Though you could walk the distance to Jupiter in a jiffy, even if you start first thing in the morning and again walk as fast as the speed of light, it will take you a tad longer to walk to the nearest stars, a few years at least. Be advised as you wander the deeps of interstellar space, the nights will seem very long. Bring a flashlight (with extra batteries), a large supply of nutrition bars and some deodorant. If you get lost, at least you won't smell bad.

Because the distance to the nearest stars is so humungous, astronomers don't bother with tiny units like kilometres and miles. When they talk about 'interstellar' distances, they talk about Astronomical units, kiloparsecs and light years – not a year with a third fewer calories, but the distance light travels in one year. Don't forget that light zips along at 300,000 kilometres (186,000 miles) per second, fast enough to beam around the world 7½ times before you can say "Holy cow that's fast!"

Yet, with the nearest stars four light years away (meaning it takes a light beam four years to travel from there to here as long as there are no delays due to road construction), you will appreciate that Earth's home galaxy which contains the Solar System and all the visible stars in the sky, is pretty darn big.

## And beyond

When you get to the edge of Earth's home galaxy, the Milky Way, just turn around and come back. It's dark out there, you will need new batteries for your flashlight, and, in all likelihood, your bills will be long overdue. Besides, the next galaxy, Andromeda, is a good two million light years away, most definitely beyond walking distance. Unfortunately, at present, there are no flights going. But check the Sunday papers for specials, the situation could change any millennium.

> **When you get to the edge of Earth's home galaxy, the Milky Way, just turn around and come back. It's dark out there.**

Bring it all back home by quoting Fred Hoyle, 'Space isn't remote at all. It's only an hour's drive away if your car could go straight upwards,' and watch the eyes of your audience go wide.

# UNDERSTANDING THE COSMOS

The problem for Earthlings is that their understanding of the cosmos is only as good as the best instruments of the day. While Galileo got by with a lens the size of a mandarin orange, today's astronomers scour the heavens using mirrors that range from the size of an average swimming pool to a new class of telescopes destined to exceed tennis court size – one of which is sited in Chile and aptly named the Very Large Telescope

> **66** The cosmos is only as good as the best instruments of the day. Galileo got by with a lens the size of a mandarin orange. **99**

(VLT). Cosmologists are happiest when stating the obvious. Another one in prospect is of football-field proportions, and known as the Overwhelmingly Large Telescope (OWL for short).

There is now a plethora of orbiting space-based observatories (e.g. Skylab). Such a location has two advantages over a ground-based observatory:

1 It allows you to star-gaze from beyond the distorting atmosphere that protects Earth – or used to, before there was a big hole in it.

2 It gives you access to new windows from which to admire and explore the Universe.

Keeping track of present and future missions (and the light they study) is nearly impossible so here are a few memorable examples. Some are named after famous scientists, like Hubble (which studies visible light), XMM-Newton (X-rays), Herschel (infrared) and Planck (microwaves); some use good words like Swift (gamma rays); some use obscure acronyms, like COROT – for Convection, Rotation and planetary Transits (visible light); and some use good words disguised as obscure acronyms, like INTEGRAL (for INTErnational Gamma Ray Astrophysics Laboratory). Thinking up others is a good gimmick but remember that specialization is all the rage now so there's no point in proposing the TOSE (Telescope tO Study Everything).

> **Pretty soon the expense and expertise of getting them up will be less than the problem of preventing them from colliding.**

To get a better understanding of objects in the neighbourhood there's nothing like dropping in for a visit, so robotic probes are sent to gather information. For example, Jupiter was first visited by Pioneer 10 in 1973, later by Pioneer 11, Voyager 1, Voyager 2, Ulysses, and the spacecraft Galileo which orbited it for eight years. Pretty soon the expense and expertise of getting them up will be less than the problem of preventing them from colliding on the orbiting hardware belt.

Should you decide to run the gamut with the present fleet of probes, you could choose from roaming the dusty surface of Mars in search of water to parachuting onto Titan, a moon of Saturn some 1.3 billion kilometres (840 million miles) away from Earth, to a rendezvous with comets – where, if you fancy a Scotch on the rocks, at least you won't be short of ice.

However, this is rocket science, after all, and things can go wrong. The Hubble trouble that sabotaged the $1.5 billion space telescope project was an error of 1.3 mm (0.05 inches), and it required a space mission to fix it. More recently, instructions using imperial instead of metric measurement sent a Mars probe into oblivion.

> 66 The Hubble trouble that sabotaged the $1.5 billion space telescope project was an error of 1.3 mm (0.05 inches). 99

To avoid further mishaps, this book will use both.

## The Big Bang

This is where everything began. Space, time, matter, energy, Belgian chocolates, left-handed scissors, ugly ties – all got started in a cosmic burp 13.7 billion years ago.

Bluffing about the Big Bang is relatively easy. You just need to know three things: it was terribly hot, stupendously dense and one day, before there were days, it began expanding. Nobody knows why. Not even the cosmologists, who spend all their time thinking about Big Things. For a blink of an eye, the entire cosmos was small enough to fit in your back pocket. Then all hell broke loose.

Because the expansion of the cosmos is a source of great confusion for people, choosing your words carefully at the beginning will save you much grief later on. To get started, it's best to repeat the following phrase to yourself before trying it out in public: The Big Bang was an explosion of space at the beginning of time. Then keep repeating it until everyone who hears it appreciates that it is truly as profound as it sounds.

**❝ For a blink of an eye, the entire cosmos was small enough to fit in your back pocket. ❞**

When you convince yourself that space and time came into existence with the Big Bang and that they didn't exist beforehand, you will have an easy time fending off such questions as: Where is the centre of the cosmos? What's the cosmos expanding into? Why is my bank account not expanding along with the cosmos? What came before the Big Bang? What does it all *really* mean?

With enquiries like these you've got two options:

1  Produce a rhetorical counter-thrust such as, "What's north of the North Pole?" followed by a shrug of the shoulders and a retreat to the bar for a drink.

2  Order a refill (and plenty of chasers) then explain what it means for space and time to come into existence with the Big Bang*.

Though the first option will score you some quick points, the second option gives you an opportunity to shine. Shine on, then. First of all, there is no centre of the cosmos. Yes, it is expanding, but that doesn't mean it's expanding from somewhere. Except for the stars, solar systems and galaxies which are all held together by gravity, and human beings who aren't on a diet, the entire cosmos is expanding. That's because SPACE ITSELF has elastic properties like rubber bands and it is SPACE ITSELF that is expand-ing, like the surface of a balloon.

> **It is SPACE ITSELF that is expanding, like the surface of a balloon.**

It will help if you speak in CAPITALS to make your point. It's okay to prod as well. "Does the surface of a balloon have a centre?" "Where is the centre of the surface of the Earth?"

Nor does the expansion of space mean it's expanding into something else. Don't spend too much time on this because you'll get nowhere. Instead, try describing the cosmos as "like a circle whose centre is everywhere but whose circumference is nowhere".

This phrase was originally meant to describe the

*Or opt out entirely by asserting, "Before the Big Bang was a Big Crunch of another universe collapsing in on itself so it should be known as the Big Bounce. In computer-speak, this is like God rebooting the system."

nature of God but it works equally well here. Indeed, *Alice in Wonderland*'s White Queen, who could believe up to six impossible things before breakfast, was probably pondering the centre and edge of the cosmos.

## A millisecond history of the cosmos

Things get off to a rousing start 13.7 billion years ago with the Big Bang. (There is much rejoicing.) Within an instant, the cosmos inflates from pathetically small to awesomely large and then resumes its normal expansion. A few minutes later, matter forms from energy. (Even more rejoicing.) Then not much happens for a few hundred million years until the first stars and galaxies form, including the Milky Way. (Sheer pandemonium.) Generations of stars wink into and out of existence until mankind's sun – a medium-sized star, as stars go – and planetary system, including Earth, form about 4.6 billion years ago. (Unbridled ecstasy.)

> **66 Less than a million years ago *Homo sapiens* make their debut. Everything pretty much evolves fine and dandy until 1970 when the Beatles break up. 99**

The first life forms appear on Earth, c. 3.8 billion years ago. (Standing ovation.) Complex life diversifies during the Cambrian Explosion some 550 million years ago, and less than a million years ago

12

*Homo sapiens* make their debut. (Congratulations all round and everyone wears a fatuous grin.) Everything pretty much evolves fine and dandy until 1970 when the Beatles break up. (Shock and dismay.)

Then the launch of the Hubble Space Telescope, in 1990, beams down a steady supply of cosmic pin-ups, the masses finally have something new to celebrate, and, as the saying goes, life goes on.

## Universal issues

Get started with a fun fact that the colour of the universe is not black or blue, or even colourless – it's beige. Think cosmic latte.

The beige universe is all there is, but the part that can be seen is described as the 'known universe'. To start any conversation with "In the known universe" is a good ploy that makes you sound smart because it hints at your awareness of an even vaster 'unknown' one (it's a salutary thought that least 96% of the universe is unaccounted for). Similarly, saying "Our universe" acknowledges there could be others next door, as it were, and that an entire ensemble of universes (formerly thought to be singular) could be floating in the cosmic imperium known as the 'multiverse'.

> **66 The colour of the universe is not black or blue, or even colourless – it's beige. Think cosmic latte. 99**

Astronomer Royal, Sir Martin Rees, is so convinced of the existence of the multiverse that he's willing to bet his dog's life on it. So that's alright then.

While the universe (known and unknown) is thought to be BIG, cosmologists don't know if it's spatially 'finite' or 'infinite'. If it's spatially finite, it's expected to be unbounded like the surface of a balloon and you'll note that the fun part of living in such a universe is the possibility of setting off in a straight line and billions of years later arriving back where you started, by which time you'll be ready for a nap. If it's spatially infinite, you can marvel that in galaxies far, far away, your other selves are having a jolly time bluffing about the spatially infinite universe.

**66** Another conundrum for cosmologists is the shape of the universe. So far the odds favour: curved like a sphere, curved like a saddle or pancake flat. **99**

Another conundrum for cosmologists is the shape of the universe. So far the odds favour:

- curved like a sphere (positive curvature)
- curved like a saddle (negative curvature)
- pancake flat (zero curvature).

Recent observations suggest that on inter-galactic scales space is flat, which, you can say, not only means that two single parallel lines will unfortunately never meet, but, more importantly, opens the cosmic door to the mysterious* dark energy, something that is easy to sound intelligent about because no-one knows what it is (see Cosmic Accounting). Besides, you can add, who wants to meet a parallel line anyway?

# CONTENTS OF THE COSMOS

## The interstellar medium

Looking up (into the Milky Way galaxy) on a starlit night, you see two things:

1 space – which astronomers call the **interstellar medium** ('medium' as in intervening substance, not as in 'Sorry, we've run out of large and small'), and
2 stars.

So when you are standing with others looking up and one of them comments "How lovely the stars are in space tonight" you can say "You mean the interstellar medium". Constant references to 'the emptiness of space' means that most people have the impression of a great big nothingness. Far from

---

* Its mysteriousness isn't officially part of the problem, but it might as well be.

15

it, you will say, the interstellar medium (**ISM** if you really want to seem an insider)) – is chock full.

What it's full of is gas (99%) and dust (1%) – the cosmic kind, remember. Though most of the gas is hydrogen, more than 100 types of molecules can be found, and you can tantalize listeners with the fact that the list includes several organic molecules important for life – some of which make up alcohol. (Now there's a word no-one would expect to be associated with this topic).

But don't get anyone's hopes up for interstellar cocktails. If your spaceship has a funnel one kilometre across and travelled through an interstellar cloud at the speed of light, it would still take a thousand years to collect enough alcohol for a single dry martini.

> **Though most of the gas is hydrogen, more than 100 types of molecules can be found, including organic molecules which make up alcohol.**

Point out that much of the gas is thinly spread, but often it gathers into towering, colourful, glowing nebulae (clumps of space dust), or into cold, dense molecular clouds. It is in these clouds that stars are born, and die. In between, they make movies and have romantic affairs (oops, wrong stars). New stars form from the debris of dead stars. The interstellar medium is the ultimate recycling bin.

## Suns and stars

The first thing you have to appreciate is that the entities seen twinkling at night are not stars but suns. Furthermore, the object we refer to as the Sun is a star not a sun. So in holding forth about them, be prepared to dazzle. Starting with the Big Picture will help, and so will the occasional dramatic pause. This is because stars (suns) are not just large, colourful whirling balls of hot gas that are held together by their own gravity, they are much more: stars, you will claim emphatically, are the reason that planets and people exist.

> **66 Contrary to appearances, stars shine 24 hours a day, 7 days a week without any statutory holidays. 99**

### Twinkle twinkle little star

Contrary to appearances, stars shine 24 hours a day, 7 days a week without any statutory holidays. Describing stars as gigantic fusion reactors should get you started. They're made mostly of hydrogen, which is lucky because that is also their main fuel for the fusion. Bingo! The energy created from this fusion is used for the shining. Altogether the process is efficient enough to give the Sun a warranty of 10 billion years, and you can relax because it has barely used half its life span. Stars shine and if they didn't, gravity would cause their collapse (it's a never-ending battle). Think of it all as a gravity-controlled nuclear bomb – a good reason not to tinker with it in your basement.

## Kaleidoscope colours and seismic sizes

Like candy in a box, stars come in a range of colours and sizes. Their surface temperatures span from hot enough to singe your hair from miles away (2,000°C) to hot enough to vapourize your space-craft, spacesuit and dental fillings (40,000 °C). You can tell which is which by their colour:

*red* – the singe-your-hair variety (which astronomers like to talk about as 'cool');

*whitish* (7,000-10,000°C) or yellowish like the Sun (6,000°C) – the type that would cause some unpleasant melting of your extremities if you ventured too close.

*blue* – the vaporize-your-spacecraft-and-its-contents kind (contrary to common parlance and bath-room tap labels, 'blue-hot' is a lot hotter than 'red-hot');

Relish the fact that these temperatures are as nothing compared with the interiors of stars which are assessed to be hotter by millions of degrees.

As for size, analogies are your best bet because not even mathematically-minded astronomers can keep the numbers straight in their heads.

Compared with, say, a head-sized Sun, a red giant swells to be the size of a camper van and a super-giant star becomes a truck. Conversely dead stars like white dwarfs and neutron stars would be as small as or smaller than the blob of ink at the end of this sentence.

## How stars are born

Everything in a star's evolution has to do with its mass (you will only say 'size' if you mean breadth). Mass determines the star's central pressure and temperature, and therefore which elements it can fuse in its core. Fear not, there is never any need to know the actual number of a star's mass because no-one understands numbers like a thousand billion billion billion kilograms anyway.

Luckily for you, the process of a star's birth is shrouded in mystery so you can say so without any loss of face. Gravity takes a large cloud and tugs it together until it is dense enough for fusion to occur in the centre. (NB: this doesn't always work. Objects called brown dwarfs are failed stars that never got hot enough to 'turn on' fusion reactions. Aw shucks.)

> **Compared with, say, a head-sized Sun, a red giant swells to be the size of a camper van.**

To generalize about the ages of stars puts you on shaky ground because, like the members of a family, they weren't all born at the same time. New stars are forming all the time, like in the Orion Nebula, and they last anywhere from millions to billions of years depending on whether they are high mass or low mass stars (you can put the dividing line at about 10 times the Sun's mass).

> **High mass stars live free and die young, while their low mass siblings plod along to collect their pensions.**

Yes, it's opposite to what you would expect – high mass stars live free and die young, while their low mass siblings plod along to collect their pensions. A high mass star of, say, 40 times the Sun's mass will burn out in a mere 3 million years. Thank your lucky stars that Earth's sun is a lightweight.

### Binaries, multiples and clusters

Why build just one when you can build a million? This seems to be the instructions in the Galaxy's Star Building Kit. Though the Sun is a solitary star, about half the stars in the Milky Way are in binary or multiple star systems, orbiting each other as they orbit the galactic centre. Some are born in clusters and just knowing two types will get you by:

**Open clusters** – a few hundred members – that are visible to the naked eye and live in the disk of the Milky Way. Their ages are typically in the tens to hundreds of millions of years. Practically new kids on the block.

**Globular clusters** – about 150 stunning swarms of a few million members, each trying to assert its unique identity. They live in the Milky Way's halo and orbit the Galactic centre in great, swooping ellipses. Likely to have formed when the galactic party was getting started about 13.6 billion years ago, they have been found to be the Universe's oldest stars – shown by 'evolutionary modelling', not their birth certificates.

## How stars die

When a star runs out of its nuclear fuel and its nights of being wild are pretty much over, the core becomes dense enough to be called 'degenerate' – and the rest of the star puffs up into a **red giant.**

When the Sun hits red giant stage about 5 billion years from now, it will swallow the orbits of Mercury, Venus, and possibly Earth and even Mars. So, the term 'giant' really does mean GIANT. Man's descendants will be toast.

> **"When the Sun hits red giant stage it will swallow the orbits of Mercury, Venus and possibly Earth."**

Eventually much of a red giant's outer 'envelope' (a gaseous shell like a glowing smoke ring that astronomers call a **planetary nebula**) blows away. You should mention that the left-over core, known as a **white dwarf**, is super-dense. Here you could mimic taking a scoop of sugar with a teaspoon, because a white dwarf is so compact that a teaspoon of its material would weigh the same as an automobile. Earth's sun (being low mass) will stay at this stage. A high mass star, on the other hand, will go out in a burst of glory – ultimately, thanks to gravity its core implodes and the whole star explodes, sending gaseous bits hither and thither – an event called a **supernova.** This star-shattering event can outshine 10 billion suns for several weeks. There's nothing like making a good last impression.

Now comes your finale. Remind others about the calcium in their bones, the water in their tissues, and the iron in their blood. Dramatic pause. Yes, you can admit, in sepulchral tones, all the atoms in our bodies were once forged in the centre of a star. Dramatic pause. We are stardust. Being able to pull this off with the right amount of gravitas is the acme of astro-bluff.

> **❝ A white dwarf is so compact that a teaspoon of its material would weigh the same as an automobile. ❞**

Denser than a white dwarf is the remnant left behind after a high mass star explodes, the **neutron star.** Display the absurdly high densities by noting that if you were able to put a teaspoon full of neutron star material on a balance scale, you would need to balance it with the combined weight of all of humanity. **Pulsars** are just neutron stars with Hollywood-style special effects added.

## Gamma ray bursts

Another cataclysmic event for Earth would be a nearby gamma ray burst – mention of which will put you on the cutting edge of cosmic expertise because their mystery is only now unfolding. You should emphasize that these brief ultra-energetic explosions absolutely out-burst all other violent phenomena in the universe. Most gamma ray bursts seem to be emissions caused by the collapse of the core of a high mass star into a black hole. Or, the merger of neutron stars.

If you want to toss out a number you can say that, over the few seconds of the burst, more energy is released than a thousand stars like the sun release in their entire lifetimes. That should make people blink.

> **" Over the few seconds of the burst, more energy is released than a thousand stars like the sun release in their entire lifetimes. "**

## Cosmic dungeons: black holes

In 1969, the following conversation was overheard between physicist John A. Wheeler and his wife:

John A. Wheeler: Have you seen my other Argyle sock?

John A. Wheeler's Wife: I forgot to tell you, the black hole got it.

John A. Wheeler: Darn! Can I get it back?

John A. Wheeler's Wife: No. It's gone through a space-time wormhole and now lives in someone else's drawer, unmatched and unloved.

John A. Wheeler: Double Darn! I liked that sock!

Since then, mysterious sock-gulping black holes have become the celebrity collapsed-star children of the cosmos. Bluffing effectively about black holes will put you in the Premier League of bluffers. Aside from a significant fraction of the population who are missing single socks, everyone loves black holes. And so far, only Wheeler (who named them), his wife, and most probably Stephen Hawking know anything about them. Unfortunately, they won't tell the rest of us in language that we can understand.

> **Aside from a significant fraction of the population who are missing single socks, everyone loves black holes.**

In fact, neither budding nor veteran astronomers understand black holes. This is why every single lecture about the cosmos since 1969 has ended with the sentences, "We're not sure, but it could be a black hole..." This is astro-speak for: "We're trying to get more telescope time at observatories in Hawaii, Chile or the Canary Islands. Even then we still won't know, but we think better with tans."

Speculation orbits around two terms: event horizon and singularity:

## 1 Event horizon

This is the trap-door-like boundary that separates the black hole from the rest of the cosmos. Here you should quickly allude to the concept of 'escape velocity', which is the speed needed to escape from the gravitational pull of an object like a planet or star. Rockets need to attain a speed of 11 kilometres (7 miles) per second to escape the clutches of Earth's gravity. In the late 18th century, John Mitchell of Cambridge and the French scientist, the Marquis de Laplace, independently suggested that you could have an object sufficiently massive (never say 'bigger') and compact that the escape velocity would equal the speed of light. Hence, nothing could escape from the object, not even light.

> **"Rockets need to attain a speed of 11 kilometres (7 miles) per second to escape the clutches of Earth's gravity."**

## 2 Singularity

When a black hole forms, all the matter is confined to a point of zero volume and so the density suddenly skyrockets to infinity. That's the singularity: a point of infinite density at the centre of the black hole. However, you shouldn't let the word 'infinite' bother you. It's a liberating word that represents a number beyond which it's impossible to go.

With these two terms at your disposal you're ready to face the masses. You can start by reiterating what you just learned about the event horizon: the gravity of a black hole is so strong that NOTHING CAN ESCAPE IT, NOT EVEN LIGHT. Shout it out to make sure it sinks in.

If asked to elaborate, you can say that a black hole is actually STRONGLY CURVED SPACE AND TIME, which behaves like an intense gravitational field. Say it without explanation and with enough conviction and if the eyes of your bluffees glaze over into utter blankness, there's a good chance you've made your point.

> **❝Imagine a bowling ball on a flat rubber sheet – it creates a deep depression. ❞**

Should you be pressed further, explain that where Newton's view of gravity was one thing tugging on another, Einstein's brilliantly warped update was that matter changed the shape of the space around itself. Here, you can quote Wheeler who put it eloquently, "Matter tells space how to curve, and space tells matter how to move." An image could help: ask your audience to imagine a bowling ball on a flat rubber sheet – it creates a deep depression. In the case of a black hole, the vortex-like depression is essentially bottomless. Advise those with vertigo not to look in.

You may also want to dispel any myths about using black holes as time machines or portals into other universes. In jumping into a black hole, you would first be stretched thin like spaghetti (if you want to sound technical, say 'spaghettified') and then crushed into oblivion. This is unlikely to make you very happy, though you may take comfort in at last being reunited with your lost socks.

## Black hole formation

As said, at the end of a high mass star's life it runs out of fuel and collapses. Sad, but there it is. Scientists have worked out if the just-formed neutron star is more than 3 times the Sun's mass, it too will collapse. Awesome, violent images like this really impress one's listeners. Then say that the remnants of the collapse would be ultra-small and compact yet still massive because the squashed matter doesn't disappear. It's already a burned out star so it's not going to start shining again. That's it: you've got the conditions needed for a black hole. And no fancy new physics was necessary.

> **66 In jumping into a black hole, you would first be stretched thin like spaghetti and then crushed into oblivion. 99**

## Black hole candidates

For extra credit, you can throw in a black hole candidate or two. They have fancy names like Cygnus X-1 or LMC X-3, where the X indicates the object is also an X-ray source. You may wonder how a black hole candidate can be an X-ray source if nothing can escape its STRONGLY CURVED SPACE, but never let on to anyone else: "It's a matter," you will say nonchalantly, "of the matter falling into the black hole that has heated up to the point where it's hot enough to emit X-rays."

> **Evidence comes from the centres of some galaxies, like the Milky Way, where it is thought that supermassive black holes are lurking.**

Even better evidence comes from the centres of some galaxies, like the Milky Way, where it is thought that supermassive black holes, with masses several million times larger than that of the Sun, are lurking. At this point it's best to adopt a serious look and say, "Look, Stephen Hawking thinks they exist," and that will silence even the most vociferous Doubting Thomas.

## Quasars

Discovered in 1963, quasars look like stars but shine a hundred times more brightly than entire galaxies (composed of, say, 100 billion stars). Found only at the farthest reaches of the universe, for many years quasars' combination of ultra-high energy and compact size presented astronomers with a juicy enigma that was politely phrased as, "What the hell are these things?"

Obviously, you should say, they're black holes.

Not your common-or-garden black hole of burned-out star fame, but supermassive black holes would be more like it. And in case you get quizzed on it, think of interstellar gas whirling toward the black hole and ultimately powering it as the gas is heated to ridiculously high temperatures on its way in.

> **It's now fashionable to speculate that most galaxies probably contain a dead quasar at their centre.**

Quasars formed early in the history of the universe, spent a good couple of billion years blazing and consuming fuel and before they knew it, their carefree glory days were over. It's now fashionable to speculate that most galaxies probably contain a dead quasar at their centre. R.I.P.

## Galaxies

From the northern hemisphere only one other galaxy, Andromeda, can be seen with the naked eye, and then only from a dark site, far from city lights. This means that more than half the people of the world will never see it because at night they are usually suffused in light – in front of the television.

Formerly known as 'island universes' (pause, for poetic effect, when you mention this), galaxies are fertile territory for bluffers, so you'll want to take in a few facts.

In small telescopes, galaxies look as fuzzy as comets (without their tails). In 1780, to assist comet hunters, the French astronomer Charles Messier published a list, sometimes known as Messier's Catalogue of Fuzzy Spots, that included 32 galaxies. These galaxies are now identified by their Messier (M) numbers; Andromeda, for example, is known among astronomers as M31.

During the early part of the 19th century, British musician and astronomer-extraordinaire Sir William Herschel, assisted by his sister Caroline and son John, identified and catalogued thousands of galaxies, extending the boundaries of the known universe.

**❝ The light from Andromeda that is received by your eyes tonight left the galaxy well before humans were still relying on rock walls to be their art galleries. ❞**

Galaxies are all at mind-blowing distances: Andromeda is 2 million light years away, so the light that is received by your eyes tonight left the galaxy well before humans were still relying on rock walls to be their art galleries. Ongoing photo searches reveal more and more, including hundreds of unusually faint galaxies, sometimes known as Crouching Giants. Presently, astronomers are leap-frogging each other to claim the most distant galaxy. You can treat these announcements with a certain circumspection because:

a   claims are sometimes retracted when they can't be verified by other astronomers;

b   the discovery will soon be superseded by another more distant galaxy. The galaxy A1689-zD1, one of the present record holders, is a whopping 13 billion light years away – it formed 700 million years after the Big Bang, just a jiffy in space terms.

Because of their abundance, most galaxies are known only by their catalogue numbers, though a few dozen have catchy names like the Whirlpool (M51), the Sombrero (M104), the Cigar (M82), the Sunflower (M63), the Black Eye Galaxy (M64), and the You Should See the Other Guy Galaxy (M65). It is now reckoned that about 100 billion galaxies populate the cosmos. These, you can say, are its true building blocks. For cinematic effect, add that if galaxies were frozen peas, they would fill the Taj Mahal.

**❝Most galaxies are known only by their catalogue numbers, though a few dozen have catchy names like the Whirlpool (M51).❞**

With all that under your belt, the next bit is a cinch.

Thanks to the American astronomer, Edwin Hubble, galaxies are grouped into three basic varieties depending on their shape: elliptical, irregular and spiral.

**Elliptical galaxies** are spherical or football-shaped, and have little gas or dust. Without any gas, they can create no new stars, so they mostly contain old stars – the Norma Desmonds of the firmament. But give them credit for being the largest galaxies in the universe.

**Irregular galaxies** (in shape, remember, not habits), like the nearby Large and Small Magellanic Clouds, visible from the southern hemisphere, are large amorphous blobs without any special characteristics so you don't need to give them a thought.

**Spiral galaxies** are the ones that resemble giant Catherine wheels of whirling stars. Seen edge on, they have a flat disk that contains a mix of stars, gas and dust and a small glowing bulge at the centre. Face-on, the full majesty of their spiral arms is evident. These are transient regions of higher density, like moving traffic jams, where clouds of gas are forming new stars.

Since spiral galaxies are so spectacular you'll be happy to hear that Earth's galaxy, the Milky Way, and the nearby Andromeda, are both spiral galaxies, with about 100 billion stars each. (Admit that it's tough to determine the shape of something from inside it, and tip your hat to the hundreds of astronomers who, over past three centuries, measured the motions of thousands of stars and nebulae to infer the Milky Way's shape.)

Steer clear, however, of the issue of how galaxies got their sizes and shapes. Blame mergers and acquisitions (like any large corporation) if you have to, or declare that the devil is in the details (a good line to use any time you're not sure). Ditto for how galaxies formed.

## The Milky Way

All the stars seen in the night sky are in Earth's home galaxy, which got its name, *Vic Lactea*, from the Romans who, lacking imagination, simply borrowed the name from a Greek myth involving the breast of a goddess spurting milk across the sky. Now that you're knee deep in it, the word 'galaxy', also has milky roots, etymologically speaking.

The Milky Way is a flying saucer-shaped disk (nearly 7,000 light-years thick) which contains the Sun and all the stars you can see in the night sky, gas and dust, a growing collection of space probes and, thanks to an orbital putt by Russian cosmonaut Mikhail Tyurin, a foam rubber golf ball.

The disk has two components:

> **“The Milky Way got its name *Via Lactea*, from the Romans who borrowed it from a Greek myth involving the breast of a goddess spurting milk across the sky. ”**

- a **spherical halo,** which contains its oldest stars;
- a **central bulge**, which contains the Galactic Centre, that is orbited by all the stars in the Galaxy.

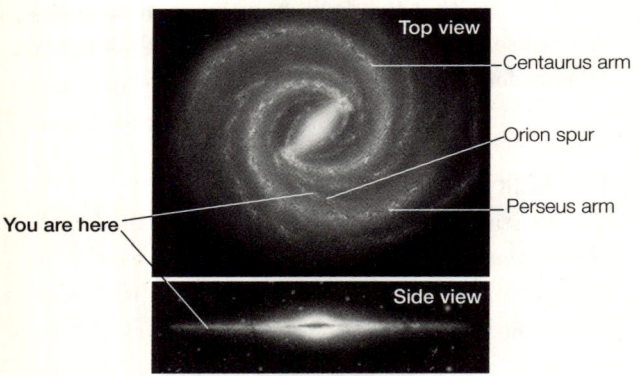

Astronomers have recently reported that the central bulge is bar-shaped, which gives you licence to paraphrase Douglas Adams: There's no restaurant at the end of the universe, but there's a bar at the centre of the galaxy.

The arms of the spiral that form the Milky Way are named after the constellations that are seen in those directions – the Perseus, the Sagittarius, the Centaurus, and the Cygnus Arms. Should anyone want to know how you know 'You Are Here', you should explain that radio telescopes detect strong sources of radio waves and surprise, surprise, Earth's sun and the centre of the Milky Way are two of these sources. So of course one can work out the whereabouts of the Earth in proportion to the Sun and the Milky Way.

From one side of the Milky Way to the other is 100,000 light years (or, in flashier astro-measurement, about 30 kiloparsecs), so if you wanted to walk across it, you should have started by now. You need to take care to avoid the black hole in the centre (most self-respecting galaxies have one nowadays) and resign yourself to missing 25,000 World Cup Final matches and, if their stars are properly aligned, at least one more England victory.

> **❝From one side of the Milky Way to the other is 100,000 light years, so if you wanted to walk across it, you should have started by now.❞**

## Closer to home

### Constellations

Bring up the stars and people think you mean the constellations which is odd when the cosmos is full of so many mind-blowing phenomena. But the patterns stars make are much loved so you might as well indulge them with a few facts before moving on to more exotic stuff.

Scattered like sailboats on the ocean, the positions of the stars and their random arrangements into constellations signify... absolutely nothing. Alas for the astrologers among us.

Astronomers use the 88 constellations to name and identify objects in various regions of space. All the heroes from Greek mythology are here, including Hercules, Pegasus, Perseus and Orion the Hunter (often recognized by his belt, seen in the winter sky). Animals are also popular, including Delphinus (which actually looks like a dolphin), Cygnus (a bit like a Swan in flight), Vela (The Snails), Volans (The Flying Fish), Monoceros (The Unicorn, about to eat the Flying Fish) and the engaging Camelopardalis (The Giraffe).

**66 Animals are also popular, including Delphinus (which actually looks like a dolphin), Vela (The Snails), Monoceros (The Unicorn)... 99**

As Earth scoots around the Sun over the course of a year, the Sun appears to track through the 12 constellations known to the ancients as the zodiac. A good bluff to make is that because Earth 'precesses' or wobbles on its axis over thousands of years, the direction of the Sun with respect to the zodiac has changed from the time the zodiac was first mapped more than two millennia ago. So point out to your audience that if they haven't stopped reading their horoscopes yet, they might consider doing so because, apart from ignoring the stupendous distances, chance alignments and the physical nature of the stars – none of which has anything to do with human affairs – astrologers are using data that's more than 2000 years out of date.

# A few famous stars

**Alpha Centauri** The Sun's nearest suburban neighbour and a triple star system. Sadly, because a light beam takes four years (actually, 4.3 years but you can let that go) to traverse the distance between it and Earth, if you could identify it (which is really only possible from the southern hemisphere, hard luck) there's not much point in dropping in to borrow a cup of sugar or the hedge clippers.

**Polaris** The North star and the tail of the Little Dipper. Erroneously thought of as the brightest in the night sky, it is, in fact, a mere 48th. Polaris just happens to be in the direction that Earth's axis is pointed and because of this, all other stars in the sky appear to revolve around it.

**Proxima Centauri** At 4.2 light years away, and 0.1 light years closer than Alpha Centauri, Proxima Centauri is the closest star to Earth after the Sun.

**Sirius** Commonly known as the 'Dog Star' (because it follows Orion the Hunter), Sirius is the brightest star in the sky. It is also twice as far away as Proxima Centauri, so plan accordingly. Sirius B, companion to Sirius, is a hot white dwarf and the most famous of its kind. It is so dense that placing a sugar cube of Sirius B material on the roof of your car would crush it.

**Zubenelgenubi** In the constellation Libra, the star whose name is the most fun to pronounce; try "zoo-BEN-el-jee-NEW-bee". It means 'the Northern claw'. Naturally, it's next to 'the Southern claw', Zubeneschemali.

# THE SOLAR SYSTEM

Within the suburbs of the Milky Way – about 26,000 light years from its centre – lies the Solar System (from the Latin *Sol* for sun) composed of a family of planets including Earth, plus asteroids and comets, all of which orbit the Sun.

> **❝Building a solar system is as easy as baking a cake, at least in theory.❞**

For a truly impressive party piece you could give a more precise location, saying: "Our Solar System is to be found in what scientists call the 'galactic habitable zone' close to the inner rim of the Orion Arm, in the Local Fluff, at a distance of 7.94 plus or minus 0.42 kpc from the Galactic Centre."

The Solar System orbits this Galactic Centre of the Milky Way and each all-expenses-paid trip takes about 225 million years. If you've been feeling tired lately, it could be this galactic jetlag.

Building a solar system is as easy as baking a cake, at least in theory. You just need loads of gas, a sprinkling of dust grains and a few million years. There is now a general agreement (which you can call 'a consensus') that it starts with a large interstellar cloud of gas and dust that shrinks down to a spinning disk as gravity pulls it all together. The centre of the disk becomes densest – and hottest – so the Sun forms here, while at various distances in the disk, dust grains collide and stick together producing larger rocky chunks called **planetesimals**. Then the planetesimals sweep up material and grow into proto-planets, and finally into planets.

The inner planets (Mercury, Venus, Earth and Mars) sweep up less material and stay small and rocky. Meanwhile, the outer ones, the gas giants (Jupiter, Saturn, Uranus and Neptune) collect more material and then pull in vast amounts of gas via their stronger gravity.

Any leftover debris of ice and rock eventually becomes dwarf planets, asteroids and comets.

Here's the whole gang pictured in their Sunday best:

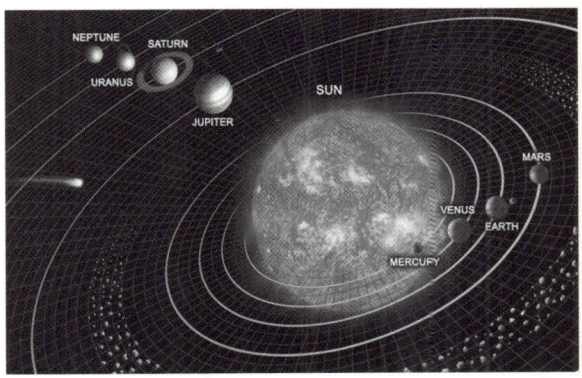

The beauty of this solar nebula-spinning disk scenario is that:

1  it gives a flat Solar System (as is observed);

2 it gets all the planets orbiting the Sun in the same direction and spinning in the same direction on their axes, while allowing for exceptions like Venus and Uranus to be explained by collisions with debris.

3 disks exist around other stars (the Hubble Space Telescope has shown them), indicating the process works elsewhere and that solar systems could be a dime a dozen.

## The age of the Solar System

Here you're in luck. Each radioactive rock gathered from Earth, from meteorites or from the Moon, is a unique clock and they all give the same answer: the Solar System is 4.6 billion years old (give or take a few million).

## The reason for sun glasses

The Sun should be your favourite star because it's so close to Earth that it lights up the sky, which relieves us from walking around in the dark all the time. Facing away from the Sun, the night sky is filled with thousands of distant suns. Enthuse about the Sun's dominance in the Solar System: it contains 99.8% of the total mass of all objects found there. And being 318 times larger than Earth, it has the potential to solve all parking problems.

You might as well extol the virtues of gravity too. Not only does it hold the Sun together – and any large 'astro-body' (great term to be used wherever possible), it compresses the gases from the relatively mild egg-frying 6000°C at the surface to a balmy egg-vaporizing 15 million degrees in the core – hot enough to stoke the thermonuclear furnace. Gravity works the same wonders on the density which, deep inside, is 13 times denser than solid lead.

No need to dwell on these stellar extremes. Just use them to acknowledge that the temperature is high enough to bake bread and to prevent electrons from binding happily with atomic nuclei so the entire gas is ionized (electrically charged) and technically known as plasma, the fourth state of matter (after solid, liquid, gas, and stuff that's been in the back of the refrigerator too long).

**66 It compresses the gases from the relatively mild egg-frying 6000°C at the surface to a balmy egg-vaporizing 15 million degrees in the core. 99**

If you're going to bluff about plasma you could strut your stuff by going on to mention one of the most feared phrases in the English language, much scarier than "I think we're out of beer" or "I just want to be friends". Brace yourself, the word is magneto-hydrodynamics. Merely whispering this can send even the bravest astrophysicists scurrying to the bar for another drink, so you should only use it if you're truly stuck. Better yet, stick to its abbreviation, MHD.

While 'magneto-hydrodynamics' doesn't exactly trip off the tongue, it means what it says regarding the study of electrically conducting fluids (plasmas) in magnetic fields. Though the business is daunting, MHD attempts to describe why there are sunspots (dark regions of lower temperature) that come and go over the 11-year sunspot cycle, and solar flares and a perpetual solar wind. It's also involved in explaining why the Sun rings like a bell, with five minute tintinnabulations. With continued observation and study, astronomers hope to make magneto-hydrodynamics as friendly sounding as "double espresso, please."

**66 MHD is also involved in explaining why the Sun rings like a bell, with five minute tintinnabulations. 99**

Connect it all to Earth with a typical space weather forecast: Sunny, with an occasional barrage of solar wind particles. These are responsible for one of the Sun's true charms, that is only visible at night. The aurora borealis ('northern lights', which also has a southern counterpart, aurora australis) is about as ephemeral and enchanting as you could hope that charged solar wind particles from the Sun zipping down Earth's magnetic field lines to reach gases in Earth's atmosphere near the poles could be.

In this display, oxygen emits green and red light,

nitrogen emits blue and a purplish red As more or fewer particles stream in from the Sun and light up different molecules, the display will appear to wave and dance. Explain that the reason the colours blaze is because the gas molecules become 'excited' (a word that's guaranteed to perk up your audience).

## The planets

The name 'planets' is derived from the Greek meaning 'wanderers' and was originally applied to the seven bright bodies that could easily be seen to change their positions in the sky. In Alexander the Great's time, the known ones were Mercury, Venus, Mars, Jupiter and Saturn. It was not long before the Romans co-opted them, as they did his territory, which is why the planets are all named after Roman gods and goddesses (e.g., Zeus, the Greek King of the Gods, was re-named Jupiter).

> **The name 'planets' is derived from the Greek meaning 'wanderers'.**

Since then, two more planets have been added: Uranus and Neptune, with a third, Pluto, that has famously come and gone. To make sure you have the order of the planets straight (starting with the one that is closest to the Sun and moving away from it), try:

My Very Educated Mouse Jumped Sideways Up auntie's Nightdress.

But if more are added to the line-up, you'll have to come up with a way to remember them yourself.

Rather than splitting the planets into "near, far and way the heck out there," astronomers divide them into three major categories:

1 **Terrestrial planets**, Mercury, Venus, Earth and Mars, which are all small and rocky like Earth;

2 **Jovian** (not 'Jovial') **planets**, Jupiter, Saturn, Uranus and Neptune, that are all gas giants like Jupiter.

3 **Dwarf planets**, the one that covers Pluto and other distant ice balls (including recently discovered Eris).

Because the planets are full of facts, sometimes referred to colloquially as factoids (not to be confused with asteroids), you do need to have a few at your disposal. Feel free to adopt the travel agent approach, offering indispensable itinerant advice which, if done with enough conviction, will give you the air of speaking from intrepid cosmic experience, rather than as an idle Internet surfer.

## Finding the planets

You can gain some invaluable bluffing points if the love of your life snuggles up to you on a dark, starlit evening, points to a bright object in the sky and asks, "What's that?" and you can say, "That's Jupiter, my love."

It could be, but it could be one of the other planets, or a star, so you need to be able to distinguish the two. Because the planets are close enough to be in walking distance (for fast walkers), they have visible disks and general-ly don't twinkle like the distant, point-like stars*. (A useful morsel is that the planets shine by reflecting sunlight, so unlike stars, their colours are not related to their surface temperatures.) The Solar System is also very flat, so if you see a steady (non-twinkling) bright object near the Sun's path in the sky, it could be a planet.

> **Planets shine by reflecting sunlight, so unlike stars, their colours are not related to their surface temperatures.**

Here's how to narrow it down.

Never say, "It's Mercury," because even seasoned astronomers have trouble picking that out of the Sun's glare.

If it's the brightest object in the sky and the time is evening or before sunrise, say, "It's Venus," (Try not to sound smug when you point out that most UFO sightings are actually Venus.)

* This is why there's no song that goes "Twinkle, twinkle little planet..."

If it's bright and reddish, you can confidently say, "It's Mars."

If it's bright and white, it's likely to be Jupiter (binoculars would confirm it with a view of the four Galilean satellites).

If it's yellowish, try "It's Saturn."

A quick glance at a reliable Internet/web site before heading outside will also improve your chances of bluffing correctly (no, that's not cheating).

## Planet distances: the scale of the Solar System

Before bluffing about the planets as physical bodies that you may or may not want to visit, it might be useful to demonstrate that a scale model of the Solar System can almost fit into most large living rooms. Naturally, it's most effective if acted out.

Take a step and describe that as the distance from the Sun to the Earth (an astronomical unit). Another half step takes you to the orbit of Mars, 3½ more will take you to Jupiter (five whole steps from the Sun), and five more bring you to the orbit of Saturn.

Nineteen steps from the Sun take you out of the living room and out of the front door to the orbit of Uranus, and Neptune would be a total of 30 steps.

The orbit of the 'dwarf planet' Pluto lies about 39

steps from the Sun – easy to recall, just think of *The 39 Steps*.

You might want to remark on the cosiness of it all, particularly because your first step was a leap over the orbits of Mercury and Venus. But be

> **66** Be aware that a trip to the outer reaches of the Sun's pull will take you 100,000 steps from the Sun. Pack a lunch. **99**

aware that a trip to the shell of comets that lie at the outer reaches of the Sun's pull – and hence the edge of the solar system – will take you something like 100,000 steps from the Sun. Pack a lunch.

Remember that as well as revolving round the Sun, planets are rotating or spinning on their own axis. One complete spin of a planet on its axis equals its day, and one full circuit of a planet around the Sun equals its year.

In your perusal of the planets it also may be helpful to know the size of each planet relative to the Earth. If you don't care for our choice of spherical objects, then find your own.

## Mercury: going to extremes

(Size: If the Earth is the size of an apple, then Mercury is size of a plum. Spins once every 59 days; orbits the Sun every 88 days.)

Named after the fleet-footed messenger of the Gods and usually lost in the glare of the Sun, Mercury gets so little attention that until the recent visit of NASA's MESSENGER only one space probe (Mariner 10) had done a flyby and that was more than 30 years ago. Then only half the surface was mapped – it looks like the Moon – and now MESSENGER is busily mapping the rest of the planet's scarred surface. Credit Mercury for the Solar System's largest crater – the formidable Beethoven; at 670 km (400 miles) wide, it would almost stretch from London to Ireland.

**66 The formidable crater Beethoven would almost stretch from London to Ireland. 99**

Advise interplanetary sojourners to be mindful of the planet's record-breaking temperature swings. The mercury on Mercury climbs to a balmy 427°C (800°F) during the day but sinks to a brisk minus 127°C (-200°F) when the Sun finally sets. Sunblock SPF 35, wool socks and long johns will come in handy.

## Venus: love is hell

(An apple a bit smaller than Earth's.) Spins (backwards) once every 243 days; orbits the Sun every 225 days – so its year is shorter than its day.)

There's no need to yell but, outside of the interior of the Sun, Venus is the MOST UNPLEASANT PLACE in the Solar System. With thick layers of cloud that insulate and cook the surface up to a tropical 480°C (900°F), Venus, named after the goddess of love, should really be named after the god of Hades. Indeed, if you can't handle the metal-melting temperatures, the crushing surface pressures 90 times that of the Earth would certainly get to you – as happened to ten Soviet probes within an hour of landing in the 1960s and '70s. Describe this migraine-inducing high pressure feeling as the same as being 1000 metres (3000 feet) under water, and advise would-be adventurers to bring lots of aspirin.

**❝Venus once contained major oceans but they evaporated thanks to the greenhouse-effect-from-hell.❞**

While mountains taller than Mt. Everest, chasms deeper than the Grand Canyon and vast lava plains from 50,000 volcanoes could impress some listeners, you're more likely to get their attention if you contend that Venus once contained major oceans but that they evaporated thanks to the greenhouse-effect-from-hell. Blame the perpetual cover of 96% carbon dioxide clouds (compared to a paltry 0.03% in Earth's atmosphere) and which include layers of hydrochloric and sulphuric acid, for ruining what could have been a nice honeymoon destination.

Speculate that the 'retrogade rotation' of Venus is due to an unfortunate accident with a wayward astro-body in the the Solar System's early 'wild party' days.

## Earth: life is good

(An apple – a green one, naturally.) Spins once every 24 hours; orbits the Sun every 365 days.)

While it may be obvious what makes Earth unique among planets of the Solar System – for example, only on Earth can you enjoy a pint of Guinness while watching the Premiere Leagues on telly – you'll need a broader outlook to put the Earth into its proper cosmic perspective.

> **66 Earth is neither too hot like Mercury or Venus, nor too cold like Mars, but instead is just right. 99**

Just remember the Goldilocks Principle. Thanks to its 150 million km (93 million miles) distance from the Sun, the temperature on Earth is neither too hot like Mercury or Venus, nor too cold like Mars, but instead is just right to allow water to exist in liquid form.

Water is essential to life as we know it – and not just for enjoying a hot bath. Seventy five percent of the Earth is covered with the liquid elixir. Water harboured the first life – in the oceans – some 3.8 billion years ago and water comprises 90% of the body weight of humans. If you're feeling glib you can wonder aloud about why our planet isn't called Water.

Point out that once life got going, it changed Earth's atmosphere because:

- Millions of years of photosynthesis by tiny single-celled creatures consumed atmospheric carbon dioxide in order to make carbon-based sugar for energy, releasing oxygen as a waste product.

- Over time, this oxygen accumulated sufficiently to give Earth the breathable atmosphere that bluffers like us depend on.

Alone among the terrestrial planets, Earth's crust is carved into plates that float on the hot mantle layer below (the jury's still out on Venus). Like any small European country, each plate has a will of its own so clashes with neighbours are inevitable, as are the ensuing earthquakes and volcanoes near the plate boundaries.

**❝ Like any small European country, each plate has a will of its own so clashes with neighbours are inevitable. ❞**

Both plate tectonics and life maintain a steady flow of carbon dioxide through the atmosphere (a thin skin only 100 km or 60 miles thick – you can say that it's proportionally thinner than the skin of an apple), which helps to moderate Earth's temperature through the greenhouse effect. If pressed on the increase of carbon dioxide in the atmosphere, assert that ending up with a nasty atmosphere like Venus would be unfortunate for all involved.

To rekindle interest after such a gloomy thought, extol the virtues of the Earth's rotation axis, whose

23½° tilt (to the plane of its orbit) results in the seasons: as Earth orbits the Sun (you can toast the 16th-century Polish astronomer Copernicus for this revolutionary idea), the northern hemisphere is tilted toward then away from the Sun for six months at a time, alternately allowing bikinis on the beach then ski weekends in the Alps. Few people are aware that Earth is actually closest to the Sun in the first week of January, the dead of winter in the northern hemisphere.

> **66 Earth is actually closest to the Sun in the first week of January, the dead of winter in the northern hemisphere. 99**

You can wrap up your Earthly bluffing with an answer to the usually rhetorical question, "Why is the sky blue?" In the rainbow of colours the Sun sends to Earth, molecules in Earth's atmosphere scatter (or collide with) blue light more than other colours – filling the sky with blue light, so you can be as terse as this: "You mean, scattered light?"

## The Moon

(If the Earth is the size of an apple, the Moon is the size of a large cherry tomato.)

Ah, the Moon (you smile) – Earth's own moon, that revolves around it (in one month) about 380,000 km (240,000 miles) distant – close enough, cosmically speaking, for a weekend rendezvous. Claimed for inspiration and solace by poets and lovelorn bluffers since time immemorial, the Moon was reclaimed in 1969 by astronauts who collected rocks and generally had a good time touring the surface before leaving behind a dune buggy and bunch of footprints.

Astronomers, who have nothing better to do in the daytime except sleep, hang around in cafés and dream up caffeine-induced theories to test at night, now think the Moon formed from the collision of a large object with the Earth in an event known, in some circles, as The Big Whack (not to be confused with a slightly bigger event known as the Big Bang). That's why the Moon always looks so sad. It's saying, "Oh my aching head."

Encourage admiration of the Moon not only for brightening dark nights but for protecting Earthlings from incoming space

> **66 Astronomers now think the Moon formed from the collision of a large object with the Earth. 99**

debris. It is also the only natural satellite in the Solar System big enough to affect its parent planet's climate, and by stabilising Earth's rotation axis, makes its climate more temperate and, over millions of years, hospitable to life. Point out that if the Moon didn't exist, humans probably wouldn't either.

Correcting a few misconceptions about man's nearest celestial neighbour should hint at the untapped depths of your cosmic knowledge. So here are the big three:

*Myth 1: The Moon can't be seen in the daytime.* "Yes it can," is really all you need to say. If challenged, say that if the Moon isn't seen at night, it must be present in the daytime sky and vice versa. Either that, or it's taking a break at the beach. It may also help to know the Moon rises almost an hour later each day.

Myth 2: *Because it presents the same sad face to the Earth, the Moon doesn't rotate.* It does, and this is easily demonstrated. Borrow the drink of a friend (preferably full) and ask her or him to waltz around you in a complete circle while keeping their eyes glued on you to make sure you don't steal a sip. Upon completion, point out that though your friend faced you through one entire revolution, she or he also faced all four walls of the room in order. Likewise, the Moon does one rotation in the time it completes one orbit around the Earth. Now you can thank your friend for donating their drink to the cause of science, and consume it.

**66 Though the Moon is involved in the tides, the Sun does half of the work. 99**

*Myth 3: The tides are controlled by the Moon.* Though the Moon is involved in the tides, the Sun does half of the work. Simply convey that the Sun pulls on one side of the Earth more than the other, creating not one but two oceanic bulges, one on either side of the

Earth. When the Moon is in the new or the full phase it makes a line with the Sun and the bulges, and high tide is even higher. Bluffing how to surf these higher waves is up to you.

The Moon is a cold, dead world (described by Buzz Aldrin, the second man on the moon, as "magnificent desolation"), and most of its craters were formed from impacts early in the history of the Solar System, long before you were born. The largest craters, all named after scientists, are up to 90 kilometres (57 miles) across, so you can imagine how New York or Paris would fare as a target.

Even larger are the basins, called 'maria' or 'seas' by Galileo when he first observed them in 1610, though he knew full well they had nothing to do with water. These too were caused by giant impacts then partially filled in by lava gurgling up from below, giving them a darker appearance, in contrast to the lighter highlands. It's the maria that make up the features of 'the man in the Moon'.

> **66 It's the basins, called 'maria' or 'seas' by Galileo, and caused by giant impacts, that make up the features of 'the man in the Moon'. 99**

Don't forget that any activity is more romantic under moonlight and whatever you're doing, it may help to know that moonlight retains a tiny but poetic fraction of the Sun's warmth.

## Eclipses

You don't need to belabour the fact that though the Sun is 400,000 times larger than the Moon, the Moon is 400,000 times closer to Earth than the Sun, which is why the two orbs appear the same size in the sky. This celestial coincidence is why we see eclipses.

If you happen to be on a pebbly beach on a moon-lit night it's a good ruse to ask those around you to select the size of pebble which, when held at arm's length, they think will cover the Moon's disk. The answer (one of the smallest) is always a surprise.

## Mooning around

People do make a case for there being other moons around Earth – for instance, the asteroid (or NEO – Near Earth Object) 2002 AA29. But, unlike the Moon, it doesn't complete an orbit of Earth; instead every 600 years it forms a bizarre horseshoe-shaped path, so it can't join Earth's elite moon club. Better known is Cruithne, the largest NEO at 5 km (3 miles) wide which was discovered in 1986. Correctly pronouncing it Croo-een-ya (it was named after an ancient Celtic tribe who occupied Britain), you will agree that it most definitely moons around since it takes 770 years per complete kidney bean configuration. However, because it orbits the Sun and not the Earth, it doesn't qualify for moon membership either.

## Mars: anybody home?

(If the earth is the size of an apple, Mars is the size of an apricot. Spins once every 25 hours; orbits the Sun every 687 days.)

Because Martians, if they exist, never seem to go out of style, Mars has always been popular – and mysterious. At the beginning of the 17th century, Johannes Kepler took a turn in the Scientific Revolution. With the data of the great naked-eye Danish astronomer Tycho Brahe (still the only astronomer to lose part of his nose in a duel and to die from drinking too much beer) vowed to solve the mystery of the shape of Mars' orbit in eight days. Eight years later, Kepler finally realized Mars and all the other planets orbit the Sun not in circles but in ellipses.

> **❝ The only astronomer to lose part of his nose in a duel and to die from drinking too much beer, Brahe vowed to solve the mystery of the shape of Mars' orbit. ❞**

Despite this knowledge, since 1960, 18 probes (4 American and 14 Russian) have failed to find the way there. Even rocket scientists have their days.

The 'red planet' (know that the colour is due to rusted iron dust and not red hot desert conditions), may present a small target for a spacecraft, but it's not without its charms. Highlights include Olympus Mons, the largest volcano in the entire Solar System, wide enough to smother England and twice as tall as the altitude that most 747s fly. There is also Valles ('Val-us') Marineris, a gorge twice as deep as the Atlantic Ocean (but caused by crustal stretching, not running water) and long enough to swallow the entire Rocky Mountain chain.

Adopt the view that water once flowed in abundance on the surface because of the plethora of dried up river channels (*canali* in Italian) and then you can thoughtfully surmise that it all evaporated away due to the ultra-low atmospheric pressure, which is one hundred times lower than Earth's. (Don't make the same mistake as American astronomer Percival Lowell who fancifully imagined them as canals after mis-translating a description by the Italian astronomer Giovanni Schiaparelli. Lowell made intricate drawings of how supposed Martians transported water down from the ice caps, and even built an observatory in Arizona to observe them.)

**66 There is also Valles Marineris, a gorge long enough to swallow the entire Rocky Mountain chain. 99**

Now it's time to play your wild card: in December 2006, NASA's Mars Global Surveyor compared photos taken four years apart of a Martian gully – and found new water deposits. Hold out the possibility that more lies just below the surface, a finding that would be welcome news for thirsty wayward astronauts.

And, just to top it off, say that scientists have now confirmed the presence of methane which may indicate microbial life underground. Remind travellers that the toxic 95% carbon dioxide atmosphere means that should they take their spacesuits off they would experience the fleeting sensation of having their blood vaporize. With an average temperature of -57°C (-71°F), it's definitely on the chilly side.

> 66 Remind travellers that the toxic atmosphere means that should they take their spacesuits off they would experience the fleeting sensation of having their blood vaporize. 99

Bobbing through the Martian sky are Phobos and Deimos (Fear and Panic, the two sons of this Roman god of War), potato-shaped moons that orbit the planet in opposite directions. Claim that they are 'captured' asteroids.

## Jupiter, by jove

(A giant, prize-winning, yellow pumpkin. Orbits the Sun once every 12 years and spins once every 10 hours which gives it the shortest day of all the planets.)

Your best approach with Jupiter is awe. Everything about Jupiter is big: its diameter is larger than 11 side-by-side Earths across its equator and it's more massive than all the planets put together, or 318 Earths if you want to get picky about it. It has earned bragging rights to spare.

Your best line is that you're a professional Cloud Watcher with a special fondness for the breadth and extremes of Jovian weather. Adjacent bands of clouds zoom around the planet in opposite directions so there's always a fresh supply of lightning, and colourful eddies and swirling hurricanes that seemingly blow for all eternity. The most obvious feature is the Great Red Spot, a perpetual cyclone wide enough to engulf two Earths, that circulates at a hair-raising 360 kilometres (225 miles) per hour on a good day.

> **❝ The planet's enormous magnetic field gives rise to lovely polar auroral displays but is a bit too hot for roasting marshmallows. ❞**

Deeper below the surface, as the Galileo probe discovered in 1995, the winds blow even harder, implying sage-like that the power comes from within. Speculate that some heat was left over from the planet's formation causing the planet to emit more energy than it receives from the Sun. Going deeper, you can surmise that Jupiter's core must be partially metallic to produce the planet's enormous magnetic field (five times stronger than the Sun's), which gives rise to lovely polar auroral displays but is a bit too hot for roasting marshmallows.

Like all the other gas giants, but ghostly in comparison, Jupiter has a thin ring, which is easily overlooked because it wasn't seen until the Voyager flybys of the late 1970s. However, what Jupiter lacks in ring substance, it more than compensates for in moons. Your line here is, "Last time I counted there were at least sixty." Most aren't worth commenting on, except the largest four discovered by Galileo (hence often called the 'Galilean satellites'), Io, Europa, Ganymede and Callisto – all named after Zeus' (called Jupiter by the Romans) favoured lovers.

> **❝ What Jupiter lacks in ring substance, it more than compensates for in moons. ❞**

You should be aware that this largest planet still has its share of mysteries. Don't let anyone try to tell you they know whether its core is solid or liquid, why the Great Red Spot is so red, or how deep the spot goes – they're bluffing.

## Saturn: a 24-carat ring

(A Halloween pumpkin, smaller than Jupiter. Spins once every 11 hours; orbits the Sun once every 29 years.)

When it comes to sheer cosmic beauty, Saturn, named after the father of Zeus, is your planet, thanks to one thing: its rings. A grand, orchestrated system, there are three main rings (A, B and C…the names being the reason why it is rumoured that scientists have to take creativity classes at weekends) detectable from Earth, and perhaps more than one thousand ringlets. Composed of dust, snowflakes, hailstones, lost airline luggage and giant abominable snowman-sized snowballs, each of the billions of ring particles asserts its individuality and orbits the planet separately.

Among the oddities of the system you can mention:

- that Saturn's rings are perhaps the flattest thing in science. Describe them as being like a sheet of paper more than 2 kilometres (1.5 miles) wide and, on the rings' scale, you've got its measure;
- that they have what appear to be spokes within them
- that the F ring, somehow (don't ask, astronomers don't know either) consists of several intertwined strands.

Apart from the rings, you can talk admirably about the planet itself. The daily forecast is simple: cloudy and windy. Saturnian winds all blow in the same direction, which keeps cyclones and hurricanes down to every 30 years or so. But even without the storms, Saturn's equator is the windiest place in the Solar System, with gusts of up to 1800 km (1100 miles) per hour. Your best traveller's advice therefore is: keep your space helmet's chin strap on.

Like Jupiter, Saturn has a mish-mash of icy, crater-pocked moons that circle the planet. Knowing a few names and traits of the 47 moons (and counting) will be quite sufficient. Try **Phoebe**, which orbits the planet backwards; **Mimas**, a spitting image of the Star Wars Death Star, or **Enceladus** (en-SELL-ah-dus), with a diameter the size of Poland and which has taken to spouting cold water geysers.

> **66** Even without the storms, Saturn's equator is the windiest place in the Solar System, with gusts of up to 1100 miles (1800 km) per hour. **99**

Save some energy for **Titan**, a moon larger than Mercury and one of only two moons in the Solar System to have an atmosphere (the other is Neptune's moon **Triton**). Titan has plenty of nitrogen, just like Earth's atmosphere. It also has methane. Light from the Sun causes some of the nitrogen and methane molecules to react together and form a stinky organic molecular smog worse than a bad day in L.A. Feel free to wonder aloud: Do Titanian (not 'Titanic') microbes exist on the moon's surface? Assure everyone that the quest is on even as you speak.

Don't forget to mention the most important thing a tourist to Saturn should bring: a rubber duck. Saturn is the only planet with an average density less than that of water so it would float. Unfortunately, finding a bathtub large enough could be a problem. Recommend improvising with an inflatable pool.

## Uranus: yours to discover

(A honeydew melon. Spins once every 17 hours; orbits the Sun every 84 years.)

Named after the Greeks' earliest supreme god, favoured pronunciations among astronomers are 'YER-uh-nus' or 'you-RAN-us' so jokes about anatomical parts should be beneath you.

Uranus is a disk-like object (instead of point-like, as is the case for stars) and was the first planet to be discovered with the telescope. When spotted in 1781 by Herschel at nearly 20 times the Earth's distance from the Sun, the size of the Solar System doubled.

**“If you arrive in the summer, you will get 21 years of daylight and in the winter, you get 21 years of deep, dark night. ”**

Uranus has a respectable number of thin dark rings (11) and an impressive bevy of moons (27 to date). But your best bluff pertains to the Uranian orbit. Because this hazy, pale blue planet is tipped on its side, where it whirls around like a cosmic Ferris wheel, day and night actually mimic its seasons in one hemisphere. So if you arrive in the summer, you will get 21 straight years of daylight and in the winter, you get 21 years of deep, dark night.

Go ahead and suggest that sometime in the past Uranus probably suffered a collision of truly cosmic proportions to give it this bizarre situation. It would be tempting to back up this collision theory by mentioning the Uranian moon Miranda, which looks as if it, too, was shattered and hastily glued back together like Humpty Dumpty. But it's probably safer to attribute the haphazard appearance of the surface to the normal geological processes – faulting, volcanism and uplift – much like those of ageing Earthlings really.

## Neptune: the new last planet

(A cantaloupe melon, smaller than Uranus. Spins once every 16 hours; orbits the Sun in no great hurry, once every 165 years.)

Neptune, named after the Roman God of the Sea, is about 50% farther away from the Sun than Uranus, but a touch more massive and a deeper shade of blue, which should give it a few extra points on your planetary beauty scale.

Most of Neptune's eight moons were discovered during Voyager 2's 1989 flyby, but you can raise a glass to the 19th-century British astronomer William Lassell, who took time between his telescope making and beer brewing to discover Neptune's largest moon Triton in 1846.

Triton stands out because it has an atmosphere (like Saturn's Titan), revolves around the planet backwards (like Saturn's Phoebe), and undergoes both geyser-like activity (similar to Saturn's Enceladus) and volcanism (like Jupiter's Io) – though due to its cold temperatures, instead of molten rock its lava is thought to be slushy ices. Rave about multi-mile high slushy plumes and ices oozing up from Triton's insides that have resurfaced this moon so that it resembles a slightly pinkish cosmic cantaloupe.

> **66** Rave about multi-mile high slushy plumes and ices oozing up from Triton's insides so it resembles a slightly pinkish cosmic cantaloupe. **99**

The cosmos is all about discovery so make sure to mention the sighting of Neptune, which was one of the great theoretical achievements – and scandals – of modern astronomy. In the mid-19th century, using a pencil and reams of paper for calculations, John Couch Adams, of Britain, and Urbain Leverrier, of France, independently predicted the position of a new planet that was disturbing Uranus' orbit. However, Adams couldn't get the Astronomer Royal to look for it and was scooped by Leverrier, who sent his predictions to a friend, Johann Galle, at the Berlin Observatory. Galle promptly found the planet within an hour of searching – at a distance of a whopping 30 astronomical units, and within a few degrees of the predicted position.

Leverrier instantly became a hero in France, much as Herschel had been in England, less than a century earlier (with his Uranus discovery). Scandal erupted when Herschel's son, John, also an astronomer, published Adams' work and claimed priority to the discovery for England. Naturally the French were infuriated. Queue trans-Channel bickering. Queue name calling. Queue enormous brouhaha. Finally, queue main characters, Adams and Leverrier who, to their credit, stayed both out of it and above it all.

The whole entertaining episode is memorialized in the names of three of five of Neptune's rings: Galle, Leverrier and Adams, and a section of the Adams ring thickens into three arcs named Liberté, Fraternité and Egalité.

> **66** Naturally the French were infuriated. Queue trans-Channel bickering. Queue name calling. Queue enormous brouhaha. **99**

## Dwarf planets

The solar system's newest category is set to grow by leaps and bounds. So far Pluto and Eris have already made the grade and they've recently been joined by egg-shaped Haumea ('hah-oo-may-a'), Makemake ('mah-kee mah-kee'), and Ceres (q.v.). Dozens more may soon follow.

**Pluto** (A blueberry. Spins once every 6 days; orbits the Sun tortoise-like, once every 248 years.)

Unless you were living in an interstellar cloud in August 2006, you will have heard that Pluto was unceremoniously removed from the planet family for not keeping its room clean. Though Pluto (from the Greek God of the Underworld) is round and orbits the Sun, two of the three criteria needed for the International Astronomical Union (an astro-body not to be messed with) to identify it as a planet, it has not 'cleared the neighbourhood around its orbit' and this is the sin that got it booted from 'planet' to 'dwarf planet'.

> **66 Pluto was unceremoniously removed from the planet family for not keeping its room clean. 99**

Had Pluto only bothered to hire a street cleaner for the thousands of ice balls in its vicinity, we could still be talking about nine planets. Instead, you'll note, it's now found to be mired in the middle of the **Kuiper Belt** – a distant, so far little-known belt of objects beyond Neptune's orbit, like the Solar System's asteroid belt but more populous. Admit that if the IAU had any sense, they would have called all these objects Kuiper (rhymes with 'viper') Belt objects, which is what they are.

Comment on Pluto's eccentricities with its orbit tilted 17° out of the pancake-flat plane of the Solar System, and its rotation axis tipped on its side like

Uranus. Then call attention to the fact that this odd little ball of rock and ice has not just one but three moons. The same collision that is likely to have knocked Pluto on its ear could have given this planet its large moon, **Charon** (after the old boatman in Greek mythology who ferried souls across the River Styx to Pluto's underworld kingdom). The two are entwined in a cosmic love embrace because Charon waltzes around Pluto once every six days while each stares moon-faced at the other. So if you land on the wrong side of Pluto, you'd never know Pluto had a companion, but on the Charon-facing side, the moon would appear frozen in the sky, never moving.

Don't moon over this *pas de deux* too long as you still need a word for **Nix** and **Hydra**, two more tiny satellites, discovered in 2005, that orbit this dwarf planet. In Greek mythology, Nix was the goddess of Night and mother of Charon, and Hydra was the nine-headed monster that nobody wanted to mess with, until Hercules put it out of everybody's misery.

> **The two are entwined in a cosmic love embrace.**

As for origins, feel free to speculate that Pluto formed out of a collection of gas and dust like the other Kuiper Belt iceballs beyond the orbit of Neptune. Don't forget the possibility that Pluto was a moon of Neptune that escaped during the same collision that knocked Triton into its backwards orbit. So there.

**Eris** (Formerly 2003 UB313, Xena and the tenth planet. Spins roughly every 8 hours.)

This icy object forced astronomers to define the word planet and consequently was excluded from the club. It's a smidge larger than Pluto but almost twice as far away. It has its own tiny moon and loops around the Sun in a leisurely 560 years.

## Way out there

Whatever other mysteries lie beyond Neptune, like the recently discovered Sedna, Varuna, Orcus, Quaoar ('kwa-war'), 2004 TY364 and 2005 RM43, you can safely wager they will also be classed as 'dwarf planets' – though it's not too late to lead a popular charge for TNOs ('trans-Neptunian objects') or something unconventional like FOIBs ('far out ice ball'), a diversionary tactic that would neatly relieve you of the obligation to know the names of any others.

## Asteroids

If large, floating rocks are your thing, you should love talking about the asteroid belt. So far more than 330,000 of these chunks of rock and metal have been tracked, numbered and named. This gives you a chance to make a name for yourself because a few million more are waiting to be catalogued.

Though their numbers are great (hence their charming nickname as 'vermin of the skies'), their sizes and masses are small. Think of them as objects that never got their act together to form a planet. Perhaps for nostalgic reasons, they are sometimes referred to as 'minor' planets. **Ceres** (the largest) is a mere 900 km (560 miles) across, a circumambulatory hike you could probably do between lunch and afternoon tea. The four most massive asteroids together comprise half of the mass of the entire ensemble.

So far, about a third of the known asteroids have orbits well enough determined to be given official numbers. Of these, only 13,000 have official names: from number 18610 (ArthurDent) to number 9621 (Michaelpalin) to number 3834 (Zappafrank). Even the Little Prince's fictional asteroid B612 has its own catalogue number (2001 TA142).

> 66 Think of them as objects that never got their act together to form a planet. 99

Most asteroids orbit the Sun between the orbits of Mars and Jupiter, but some whiz near Earth's orbit and could cause trouble. These NEAs ('near-Earth' asteroids), and there may be a couple thousand that are larger than a kilometre in diameter. One was a bit larger still, Asteroid No. 4581 that passed within 700,000 km (400,000 miles) of the Earth. Before someone says, "Pah! that's still twice as far away as the Moon," point out that it passed through the exact location that Earth occupied only six hours before.

Privately, however, you know that vigilance is improving and scientists are actively working on ways to deflect any strays, so don't cash in your life savings just yet.

## Meteors and meteorites

Shooting stars, falling stars and meteors have nothing to do with stars, so you need to clear that up right away. Instead, they are mere specks of comet dust that burn up as they whiz through the atmosphere (a bit like tyre skidmarks). If you wish on one, try wishing they would last a little longer.

> **66** They are mere specks of comet dust that burn up as they whiz through the atmosphere (a bit like tyre skidmarks). **99**

Seeing dozens of meteors in a single night has nothing to do with the Second Coming or the Apocalypse. More likely it's a meteor shower and occurs several nights of the year (for instance, August 12 and December 13) as Earth passes through the remnant dusty tail of a comet.

Meanwhile, meteorites are rocky and metallic chunks of asteroids big enough to survive the burning trip through the atmosphere and crash to Earth's surface. More than 150 have been large enough to cause an impact crater. Many of these impacts would blast impact debris up into the atmosphere, blocking the Sun and putting everybody in a bad mood. It could ruin your picnic.

## Comets

Once thought to be harbingers of doom, comets are floating icebergs that occasionally get tugged towards the Sun. They have an icy head with trapped gases and dust, and a tail that grows as it nears the Sun (giving rise to their being known, historically, as 'long-haired' or 'bearded' stars). One thing to point out, literally, is that from a photograph you can't tell the direction a comet is travelling to because the tail always points away from the Sun whether the comet is coming or going.

If you missed Halley's Comet in 1986, the next date to miss will be 2061.

# LOOSE ENDS

## Evidence for the Big Bang

You can thank Sir Fred Hoyle, who hated the idea of the universe having a beginning, for coining, in a flippant way, the name Big Bang. But as a theory (in the sense of a detailed framework, not in the

73

sense of a hypothesis), the Big Bang is in pretty good shape, outlasting all other contenders, which is a good enough reason for cosmologists to take it seriously. If it's a good scientific theory, it must have observational evidence (note that scientists seek 'evidence' rather than 'proof') in its favour. Three pieces of evidence should suffice:

## 1 Hubble's law

Everywhere you look, distant galaxies are speeding away from us (and each other) more rapidly than nearby galaxies are. This is known as Hubble's Law, after Edwin Hubble, who, in turn, built on the pioneering work of the cosmically named Vesto Slipher.

**❝There must have been a time when everything was right on top of each other.❞**

Hubble was reluctant to accept the implication of his work – that space is expanding – so it will be easier to convince others if you think about the situation in reverse: at earlier and earlier times, all the galaxies were closer and closer together. So, there must have been a time when everything was right on top of each other. This time is the moment of the Big Bang, when all of space and time exploded into existence.

Keep in mind that because SPACE ITSELF is expanding, and driving the galaxies apart, the Hubble Law does not imply that we are at the centre, because any other observer in any other galaxy would think themselves to be the focal point.

## 2 Cosmic microwave background

Just as when taking your dinner out of the oven it takes a while to cool off, the universe has been cooling down since its ultra-hot glory days when all matter and energy were squished together like peas in a micro-pod. In the 1950s, science popularizer, George Gamow predicted it should be possible to measure the temperature of this leftover radiation.

Sadly for him he published his work in an obscure journal and no-one was aware of it until in 1965 American radio astronomers Arno Penzias and Robert Wilson serendipitously measured the temperature of the cosmos.

With an enormous microwave receiver that looked like the ear of an old phonograph player, they discovered that no matter which direction of the sky they pointed to they picked up a tiny background signal.

> **The universe has been cooling down since its ultra-hot glory days when all matter and energy were squished together like peas in a micro-pod.**

It was the relic radiation from the Big Bang itself and had a temperature of a cool 3° above absolute zero (about -270°C). You can brag that the satellites COBE ('Koe-Bee') and WMAP have updated and verified this work to staggering precision. A few percent of the TV 'snow' flicking between channels are caused by the background radiation of the universe.

### 3 Cosmic helium abundance

The third piece of evidence is known as the cosmic helium abundance and it also involves George Gamow. Gamow was trying to determine where all the elements in the periodic table came from. He decided the early universe was hot enough to create them and went about doing the calculations. He was right in the idea, but wrong in the details.

> **What you see isn't always what you get when it comes to galaxies and the universe. Things just don't add up.**

Nevertheless, you should know that his initial work was with Ralph Alpher, and Gamow, being the jokester that he was, added Hans Bethe's name to his paper on the origins of matter so that the three authors would be Alpher, Bethe and Gamow, a pun on the first three letters of the Greek alphabet. Later calculations by others, in punless papers, correctly predicted the helium abundance in the cosmos.

## Cosmic accounting

What you see isn't always what you get when it comes to galaxies and the universe. Things just don't add up. There are two methods of tallying the evidence, both of them dark, though each has its allure. Dark matter relates to weighing individual galaxies (and clusters of galaxies, but that's a detail) and the mysterious dark energy relates to the entire universe. The former is a minor accounting glitch and the latter is an egregious error of the sort hushed up by politicians.

## Dark matter

Whichever way one sums up the mass of galaxies the maths do not agree. Galaxies seem to contain ten times more dark matter (causing things to move faster than they should) than visible matter.

The two main types of dark matter are:

**1 MACHOS** (massive compact halo objects) – planet-sized objects that might include failed and dead stars (don't forget to say, "Astronomers aren't sure, but they could be black holes") that live in the spherical halo of the Milky Way, and presumably other galaxies too. They take up about 20% of galactic dark matter.

**2 WIMPS** (weakly interacting massive particles) – that take up the other 80% and are thought to be exotic (unknown) subatomic particles that barely interact with ordinary matter. Millions of WIMPS are likely zipping through your body at this moment. If that doesn't give you a buzz, at least it inspires the physicists who are now building detectors to catch them (out there, of course, not in humans, but you could keep people in the dark for a moment or two). For bonus points, speculate that this kind of dark matter is normal matter in a parallel universe.

## Mysterious dark energy

Dark energy is inferred from accounting based not on mass, but on the distances of distant galaxies and the curvature of space.

Measurement of the distances to distant galaxies has recently shown that the universe isn't just expanding, it's accelerating. While that's cool in and of itself, even cooler is that Einstein allowed such a possibility in his equations. (For technical reasons, he referred to it as his 'biggest blunder' and because it looks to be a hot topic in cosmology for years to come, your line is, "Would that I could make such a blunder myself.")

> **❝ Measurement of the distances to distant galaxies has recently shown that the universe isn't just expanding, it's accelerating. ❞**

More thrilling still is the evidence from measuring small temperature differences in adjacent regions of space – the cosmic microwave background. These allow a measure of the overall curvature of space. Tarry a bit here to remind listeners that the curvature of the universe could be positive, negative or flat. From the temperature fluctuations, the curvature is found to be close to zero.

That is to say, space appears to be flat.

However, cosmic accounting has shown that the total amount of matter in the universe (both visible and dark) isn't enough to explain why space is flat. Now you can gloat about the fact that the amount of dark energy required to explain the acceleration of the universe is just enough to explain the flatness of space. What could be more balanced?

As nobody knows what the mysterious dark energy is, your

> **66 Think of them as objects that never got their act together to form a planet. 99**

pièce de résistance is simply to add up the contents of the universe (allow a few percent error):

4% visible matter,

22% dark matter, and

74% mysterious dark energy.

This ledger is perhaps the crowning achievement of 20th-century cosmology.

It's no surprise that the nature of dark energy is a mystery. If you want to go out on a limb, you've got three options:

1   You could say it's the repulsion of space itself (technically known as 'Einstein's cosmological constant).

2   You could say it's due to an all-pervading fluid that acts repulsively (technically known as 'quintessence').

3 You could say that Newton and Einstein were wrong about gravity (technically, this isn't advisable, but it's still possible).

At a pinch, you could invoke string theory, but as that's even more mysterious than dark energy, this ploy is likely only to make string theorists happy. Whatever you decide, acknowledge that cosmologists are burning the midnight oil trying to figure it out. Remember too that you can always incorporate new information with a knowing nod and with Alice's catch-all, 'Curiouser and curiouser'.

## Is anybody out there?

You will undoubtedly want to bring up the greatest discovery the human race has yet to make (besides belly button lint): evidence for extra-terrestrial life. So far, the evidence – at its most optimistic – can be described as somewhere between non-existent and inconclusive. It's best to take a two-pronged approach here:

1 Divulge the fact that astronomers have now found more than 300 planets around other stars. This is a perfectly respectable number even if (because they are the easiest to find) most of those discovered so far are Jupiter-sized or larger. Maintain a stoic confidence that new breeds of satellite telescopes are sure to find something more Earth-like.

2 Cite a famous meteorite, and perhaps even get to know it by name: ALH84001. Named after the Allan Hills of Antarctica, where it was found, this stray rock dates from 4.5 billion years ago – the birth of the Solar System – but via a complicated series of events is actually thought to come from Mars. The idea is not so far-fetched. Start with the fact that the rock is chemically similar to Martian soil and then mutter about something bashing into Mars millions of years ago, sending Mars-chunks of debris into space where they floated around the Sun for millions of years more until a few dropped in on Antarctica.

**66 They floated around the Sun for millions of years more until a few dropped in on Antarctica. 99**

However, your main attraction here is the contents of ALH84001. Wax poetical about the humble origins of life as lowly bacteria. Ponder aloud about what an end to our cosmic solitude might mean for universal peace, brotherhood and afternoon tea. And introduce the evidence: tiny, tubular and egg-shaped microfossils, similar to Earth bacteria though much, much smaller, that have been discovered within ALH84001, as well as long crystal chains that could only be created by once-living things.

Rather than allowing that ALH84001 was contaminated by life on Earth, hold out the possibility that life here was seeded by such rocks. Bolster your case with meteorites that have crash-landed with a fiesta of amino acids (chemical building blocks for life), including some not found on Earth.

> **66 Meteorites have crash-landed with a fiesta of amino acids (chemical building blocks for life), including some not found on Earth. 99**

Beyond the non-existent and inconclusive, you could also remark upon the tantalizing. Probes continue to scour among the moons of the Jovian planets (such as Europa, Titan, Enceladus and Triton), not for someone to invite home for a beer, but for something that could possibly wink at you from under a microscope.

Finish this little E.T. speculation by advising your listeners that whether life exists elsewhere or not, both possibilities are equally profound.

## The fate of the universe

The physicist Niels Bohr said, 'Prediction is difficult, especially about the future', but it's not only possible to bluff about the ultimate fate of the cosmos, it's also easy. There are three strong contenders:

1 That gravity will ultimately halt the current expansion and order all the galaxies to return to where they started from, causing a big pile-up.

2  That after an infinite length of time the cosmos runs out of energy and just stops expanding.

3  That it goes on growing, and galaxies keep receding from each other for ever and ever amen.

To make it easier on yourself, think of these options as the Big Crunch, the Big Vacuum and the Big In-Between. These too have to do with the curvature of space. Don't get bogged down here. It's the Big In-Between that corresponds to a flat universe, and if that's right – and the universe is accelerating – it will keep expanding until the end of time. Queue music fading gently into the good night.

## GLOSSARY

**Astronomical unit**  The distance of the Earth from the Sun (150 million kilometres, 93 million miles or 8 light minutes). By anyone's reckoning, astronomical.

**Big Bang**  The event that started this great and wonderful mess we're in. Not to be confused with sex.

**Black hole**  Deep, dark dropout in the fabric of spacetime sometimes known as the tunnel at the end of light.

**Constellation** Fanciful connect-the-dot design in the sky invented by the ancients that have yet to be bought, trade-marked or corporatized.

**Cosmic background radiation** Radiation hang-over induced by the Big Bang party.

**Comet** Dirty snowball that orbits the Sun; name for one of Santa's reindeer.

**Cosmologist** Bigger version of an Astronomer

**Galaxy** City of dust, gas and stars, free of traffic jams yet susceptible to occasional violent activity; open all night.

**Gravity** A pull (no strings attached), thought by Newton to be due to the forces between masses and by Einstein to be due to the curvature of space.

**Greenhouse effect** Warming of an atmosphere due to trapping of incoming solar radiation by atmospheric gases. Like most things, best in moderation.

**Halo** Spheroidal envelope that surrounds a spiral galaxy and blesses it.

**Interstellar medium** Whole lotta gas, spread thin, with a sprinkling of dust and possibly an alien spaceship or two found in the space between stars. (Between galaxies, you'd find the intergalactic medium, which is gas spread even thinner.)

**Kiloparsec (kpc)** One thousand parsecs, or 3,262 light years. You can kpc why the term is quite useful.

**Light-year** The distance light travels in one year, without stopping for fuel, coffee or bathroom breaks.

**Local fluff** Thin interstellar cloud that the Solar System is currently moving through. No known date for dispersal.

**Meteor** A wish-inducing piece of comet dust that buzzes – and blazes – through the atmosphere. The ultimate fly-by-night operation.

**Meteorite** Wayward chunk of space rock that would ruin your day if it hit you.

**Moon** A solar body that orbits a planet; inspiration for moonshine.

**Orion's Belt**, Tie, Socks, Top Hat and Overcoat. Constellation easily seen in the winter time.

**Planet** Home-base for terrestrials that's not large enough to be self-luminous.

**Pulsars** Compact stellar remnant that can spin ultra-fast without getting dizzy.

**Quasar** Cosmic celebrity known for once being super-bright (and super-far) and now mostly super-faded.

**Red Giant** Aged star with a swelled head.

**Singularity** Point of infinite density thought to exist at the centre of a black hole and at the beginning of the universe; closed to visitors who value their lives.

**Solar wind** A steady stream of charged particles emitted by the sun, less politely known as solar flatulence.

**Spaceliners** Virgin's rocket planes, set to lift off around 2010. Book now to accompany celebs such as Brad Pitt and Sigourney Weaver. Be prepared for a week's training for the 3 hour flight.

**Spacetime** Flexible, inseparable pair of space and time like salt and pepper, beer and pizza, and night and day.

**Star** a) glowing ball of gas that either burps itself out of existence or fades away; (b) what you will be with proper use of this book.

# THE AUTHOR

Daniel Hudon's interest in the cosmos has been expanding since he was a kid (in Calgary, Canada) looking at the sky with binoculars, an activity that, even after obtaining multiple post-secondary degrees in the subject, he still considers one of life's great joys, right up there with world travel and World Cup soccer.

He admits to spending, on his first trip to the 3.6 metre (11ft 8in) Canada-France-Hawaii Telescope, almost as much time outside on the catwalk around the dome observing the dark starlit sky, as inside observing the telescope monitors. So, of course, the prospective Very Very Large Telescope is out of bounds as he might never come home.

When not promoting the universe at Boston University, he tries to plumb the dark matter of his mind for articles in children's science magazines about things that go bump in the cosmic night.

### Quantum Universe

Never commit yourself about the quantum realm even to a 'probably'. Anything you declare to be 'probably true' could return to haunt you and, it can be said with confidence, probably will. If you know what's good for you, a 'possibly' is the farthest you will go.

### Genetics:

It is not that scientists are so terribly fascinated with peas and fruit flies, but rather that the peas and flies have a life cycle of a few weeks, thousands can be kept in the lab for very low cost, and they have similarities (at least from a genetics' perspective) to humans.

### Golf:

Never be foolish enough to play with someone who is a lot better than you are: it is courting disaster. The more you try to match their long hits, the shorter and more desperately wayward yours become. If you are the worst player in a foursome you can become a manic depressive in no time at all.

**Rocket Science:**

Contrary to popular belief, rockets and satellites are different things. The analogy could be made between pregnancy and childbirth. It takes approximately nine months between the consummation of a launch agreement and the actual lift-off.

**Philosophy:**

Of course, any sensible theory is neither one thing nor the other; and it is generally safe to say something to that effect without fear of having to say just how much of one, or the exact proportion of the other.

**Jazz:**

It's a pretty safe bet, if an unfamiliar 1940s/50s name crops up in conversation, that you can get away with saying: 'Oh, yes. Parker roomed with him for a time.' Even if he didn't, nobody is going to be sure enough to say so. And it somehow makes the unknown musician seem a better player.

Comments on other titles

## On the series:
"Chock full of the basic information that's needed to pass yourself off as knowledgeable." *The Globe and Mail*

## Management:
"Just what a Bluffer's Guide should be: short and concise, yet offering very reasonable advice to those who aspire to become good managers." Reader from Switzerland

## Public speaking:
"Get tips from the experts plus how to act with unexpected noise, power failure, etc while you are speaking. Very funny and very enjoyable." Reader from The Netherlands

## Whisky:
"Everything you need to know about whisky. Light-hearted in style and full of useful facts. Highly recommended." Reader from Scotland

## Marketing:
"Any marketing person who has not read this book has almost certainly wasted their time and money reading all the others. It's funny, witty, and true." Reader from London

# the Bluffer's® Guides

Oval Books

*This Bluffer's® Guide is available as a downloadable audiobook: **www.audible.co.uk/bluffers**

We like to hear from our readers.
Please send us your views on our books
and we will publish them as appropriate on
our web site: ovalbooks.com.

Oval Books also publish the best-selling
Xenophobe's Guide® series –
see www.ovalbooks.com

Both series can be bought via Amazon or directly
from us, Oval Books through our web site
www.ovalbooks.com or by contacting us.

Oval Books charges the full cover price
for its books (because they're worth it) and
£2.00 for postage and packing on the first
book. Buy a second book or more and postage
and packing will be entirely FREE.

To order by post please fill out the accompanying
order form and send to:
Oval Books
5 St John's Buildings
Canterbury Crescent
London SW9 7QH

cheques should be made payable to: Oval Books

or phone us on +44 (0)20 7733 8585
or visit our web site at: www.ovalbooks.com

Payment may be made by Visa or Mastercard and orders are
dispatched as soon as the card details and mailing address are
received. If the mailing address is not the same as the card holder's
address it is necessary to give both.

Oval Books